Chief Security Officer (CSO) Guideline

1.0	Title	3
2.0	Revision History	3
3.0	Commission Members	3
4.0	Committee Members	3
5.0	Acknowledgment	4
6.0	Guidelines Designation	4
7.0	Scope	4
8.0	Summary of Guideline	4
9.0	Purpose	4
10.0	Overview	4
11.0	Reporting Relationship	5
12.0	Model Function	5
	12.1 Model Profile Diagram	6
13.0	Key Responsibilities and Accountabilities	7
	13.1 Key Success Factors	7
	13.2 Strategy Development	7
	13.3 Information Gathering and Risk Assessment	8
	13.4 Organization Preparedness	8
	13.5 Incident Prevention	8
	13.6 Securing People, Core Business, Information, and Reputation	8
	13.7 Incident Response, Management, and Recovery	9
	13.8 Investor Relations, Public Affairs, and Government Relations Coordination	9
14.0	Key Competencies	9
15.0	Experience	11
16.0	Education	11
17.0	Compensation	11
APPENDIX A		13

Copyright © 2004 by ASIS International

ISBN 1-887056-52-1

ASIS International (ASIS) disclaims liability for any personal injury, property, or other damages of any nature whatsoever, whether special, indirect, consequential, or compensatory, directly or indirectly resulting from the publication, use of, or reliance on this document. In issuing and making this document available, ASIS is not undertaking to render professional or other services for or on behalf of any person or entity. Nor is ASIS undertaking to perform any duty owed by any person or entity to someone else. Anyone using this document should rely on his or her own independent judgment or, as appropriate, seek the advice of a competent professional in determining the exercise of reasonable care in any given circumstance.

All rights reserved. No part of this publication may be reproduced, stored in a retrieval system, or transmitted, in any form or by any means, electronic, mechanical, photocopying, recording, or otherwise, without the prior written consent of the copyright owner.

Printed in the United States of America

10 9 8 7 6 5 4 3 2 1

Chief Security Officer (CSO) Guideline

1.0 TITLE

The title of this document is the Chief Security Officer (CSO) Guideline.

2.0 REVISION HISTORY

Baseline Document

3.0 COMMISSION MEMBERS

Sean A. Ahrens, CPP, Schirmer Engineering
Norman D. Bates, Esq., Liability Consultants, Inc.
Regis W. Becker, CPP, PPG Industries
Jerry J. Brennan, Security Management Resources, Inc.
Chad Callaghan, CPP, Marriott International, Inc.
Pamela A. Collins, Ed.D., CFE, Eastern Kentucky University
Michael A. Crane, CPP, IPC International Corporation
Edward J. Flynn, CFE, Protiviti, Inc.
F. Mark Geraci, CPP, Bristol-Myers Squibb Co.
L. E. Mattice, Boston Scientific Corp.
Basil J. Steele, CPP, Sandia National Laboratories
Don W. Walker, CPP, Securitas Security Services USA, Inc.

4.0 COMMITTEE MEMBERS

Donald P. Bitner, Amgen
Keith D. Blakemore, CPP, Boise Office Solutions
Jerry J. Brennan, Security Management Resources, Inc.
David Burrill, British American Tobacco
George K. Campbell, Security Risk Management Consultancy
John C. Cholewa III, CPP, Sprint Corp.
Grant R. Crabtree, CPP, ALLTEL Corp.
Joseph A. DiDona, Reader's Digest Association, Inc.
Robert F. Fox, Sprint Corp.
Timothy L. Gladura, Cardinal Health, Inc.
John Hartmann, The Home Depot
Robert W. Hayes, CPP, CFE, Business Security Advisory Group, LLC
Michael A. Howard, Microsoft Corporation
Don L. Hubbard, PricewaterhouseCoopers LLP
Mark S. Lex, CPP, Abbott Laboratories
Robert F. Littlejohn, CPP, Avon Products Inc.
John E. McClurg, Lucent Technologies, Inc.
Raymond A. Mislock, Jr., E. I. duPont de Nemours & Co.
David R. Saenz, Levi Strauss & Co.
Victor E. Thuotte, Jr., Fidelity Investments
Reginald J. Williams, CPP, The Boeing Company
Timothy L. Williams, CPP, Nortel Networks
W. Lance Wright, USEC Inc.

Chief Security Officer (CSO) Guideline

5.0 ACKNOWLEDGMENT

ASIS International would like to express its foremost appreciation to W. Lance Wright, Vice President Human Resources & Administration, USEC Inc., author of the original Chief Security Officer (CSO) white paper presented at the 2002 ASIS Annual Seminar and Exhibits, for his time, assistance, and use of material, which contributed significantly to the development of this Guideline.

6.0 GUIDELINES DESIGNATION

This Guideline is designated as ASIS GDL CSO 06 2004.

7.0 SCOPE

The Chief Security Officer (CSO) Guideline has applicability in both the private and public sector environments, which must evaluate and respond to the continuingly increasing and changing risks to their assets and organizations both domestically and globally.

8.0 SUMMARY OF GUIDELINE

The CSO Guideline is designed as a tool to allow an organization to decide upon and provide a security architecture characterized by appropriate awareness, prevention, preparedness, and response to changes in threat conditions. This Guideline is structured at a high level, although specific considerations and responses also are addressed for consideration by individual organizations based on specific risk assessment and requirements.

9.0 PURPOSE

ASIS International has developed the CSO Guideline as a model for organizations to utilize in the development of a leadership function to provide a comprehensive, integrated security risk strategy to contribute to the viability and success of the organization.

10.0 OVERVIEW

Today's business risk environments have become increasingly more severe, complex, and interdependent, both domestically and globally. The effective management of these environments is a fundamental requirement of business. Boards of Directors, shareholders, key stakeholders, and the public correctly expect organizations to identify and anticipate areas of risk and set in place a cohesive strategy across all functions to mitigate or reduce those risks. In addition, there is an expectation that management will respond in a highly effective manner to those events and incidents that threaten the assets of the organization. A proactive strategy for mitigation of the risk of loss ultimately provides a positive impact to profitability and is an organizational governance responsibility of senior management and governing boards.

It is the thesis of this Guideline that the skills and competencies essential to active protection and measurably effective response to the modern threat environment are far more critical than ever before. Effective leadership within the top levels of the organization and its related security functions are imperative. Organizational reputation, the uninterrupted reliability of the technical infrastructure and normal business processes, protection of physical and financial assets, the safety of employees, and shareholder confidence all rely in some measure upon the effectiveness of an accountable senior security executive.

Traditionally, what has previously been lacking is a single position at the senior governance level having the responsibility for crafting, influencing, and directing an organization-wide protection strategy. In many organizations, accountability is dispersed, possibly among several managers in different departments, with potentially conflicting objectives.

The diversity of today's risks comes in a complex matrix of interrelated threats, vulnerabilities, and impacts, the safeguards for which must therefore be interdependent. The ability to influence business strategy and address matters of internal risk exposure requires a Chief Security Officer (CSO) at the appropriate level in the organization. This should be achieved through the restructuring and focusing of current efforts through this single senior management function, eliminating the redundant and narrow interests that may be present in vertical departmental structures.

11.0 REPORTING RELATIONSHIP

Appropriate reporting relationship decisions for the CSO position should be driven by an evaluation of the current organizational structure. It is strongly recommended that the position report to the most senior level executive of the organization, to ensure for a strong liaison with the Board of Directors and its operating committees. The relative position in the hierarchy is a signal not only of top management's commitment and support, but also, ultimately, of the legitimacy imputed to the security program.

12.0 MODEL FUNCTION

The diagram following this section lays out the scope of an organization's protection program that includes functional areas of responsibility, key processes, and discussion of work elements that may be found within the organization. While "ownership" in a strict sense is not essential, strategic accountability and effective influence is. Leadership may take the form of a Security Council or actual managerial and budgetary accountability for all security functions. The culture and business model at work within the organization will guide decisions seeking to establish the best approach. But the concept of an organizational vision and voice for the protection mission is at the heart of this Guideline.

It is recognized that many different approaches may be taken to align with the host organizational model. To aid in understanding and facilitating implementation, this Guideline presents a model position description (Appendix A).

Chief Security Officer (CSO) Guideline

12.1 Model Profile of a Chief Security Officer Function

Risks	Potential Processes & Services	Skill Set Required
Human Resources & Intellectual Assets	Global Security Policy & Procedures Administration	**Relationship Manager** *Develops, influences and nurtures trust-based relationships with business unit leaders, government officials and professional organizations. Acts as a consultant to all organizational clients.*
Ethics & Reputation	Technology & Infrastructure Protection	
Financial Assets		
Information Technology (IT) Systems	Information Risk Management	**Executive Management & Leadership** *Builds, motivates and leads a professional team attuned to organizational culture, responsive to business needs and committed to integrity and excellence.*
Transportation, Distribution & Supply Chain	Business Continuity, Crisis Management & Response	
Legal, Regulatory & General Counsel	Employee Risk Awareness	
Physical & Premises	Investigative & Forensic Services	**Subject Matter Expert** *Provides or sees to the provision of technical expertise appropriate to knowledge of risk and the cost-effective delivery of essential security services.*
Environmental, Health & Safety **	Safe & Secure Workplace Operations	
	Tailored Business-Process Safeguards	**Governance Team Member** *Provides intellectual leadership and active support to the organization's governance team to ensure risks are made known to senior management and the Board.*
	Insurance & Risk Transfer	
	Risk Assessment, Evaluation & Testing	**Risk Manager** *Identifies, analyzes and communicates on business and security-related risks to the organization.*
	Executive Protection	
	Background & Due Diligence Investigations	**Strategist** *Develops global security strategy keyed to likely risks and in collaboration with organization's stakeholders.*
	Business Conduct & Security Compliance	**Creative Problem Solver** *Aids competitiveness and adds value by enabling the organization to engage in business processes to mitigate risk. Acts as a positive change agent on behalf of organizational protection.*
	External & Government Relations	
	Business Intelligence & Counter-Intelligence Support	

** Recognizing that EH&S may be structured outside the scope of security functions, there are still significant risk issues to an organization. Since many organizations have combined their EH&S and security functions, we have chosen to present it in this Guideline for consideration.

13.0 KEY RESPONSIBILITIES AND ACCOUNTABILITIES

The CSO is a full partner in the governance infrastructure of the organization. If a comprehensive assessment of any of those areas of risk, noted in the above model, supported the need for a function specific security role, the assignment of high-level accountability will better ensure an integrated security strategy with less duplication and lower cost.

A core responsibility is the management of effective working relationships among client groups. This Guideline recommends that front-line accountability for protecting the business should fall to the managers of each operating unit with the appropriate organization's security function providing the risk assessment, policy, and supporting infrastructure.

This model requires a senior executive that can lead and enable the line businesses in the implementation of policies and the detection and reporting of risk in a timely manner. Being an effective business process enabler will require the incumbent to be a creative problem solver, a leader who can blend common sense controls with efficient and productive business processes.

It is also necessary that the incumbent bring subject matter expertise to the position. Leadership of a multi-faceted security program requires generalist knowledge, but it is likely that he/she will have come from a background within the business, a governance function, or some element of the security mission. Credibility within the team and the vision to craft an integrated strategy depends on the CSO's ability to understand, value, and articulate the varied security missions.

13.1 Key Success Factors

- Ability to build sustainable competitive advantages through pragmatic, innovative security solutions.
- Demonstrated integrity and ability to maintain principles under internal and/or external pressure.
- High-quality analytical skills, management experience, and exceptional relationship management competencies.
- Qualitative experience in strategic planning and/or policy development at a senior level.
- Ability to anticipate, influence, and assist the organization to assess and rapidly adjust to changing conditions and trends (internal and external) of importance to the direction of the organization.
- Effectiveness in communicating recommended courses of action for innovative, business-oriented responses.
- Passion for excellence and a demonstrable orientation toward successful staff development.

13.2 Strategy Development

A key responsibility of the CSO is to develop and implement a strategy that demonstrates the processes in understanding the nature and probability of catastrophic and significant security risk events. The strategy must outline in detail the plans to prevent and prepare for an adverse event, including state-of-the-art awareness, training, exercises, and methodologies to inculcate contemporary security programs and processes throughout the organization. The strategy also should cover continuity of business operations from any security-related attack or catastrophic event. The CSO must be capable of clearly communicating this strategy, costs, and related impacts to the highest levels of the organization and the Board of Directors and its operating committees.

13.3 Information Gathering and Risk Assessment

The CSO is responsible and accountable for gathering and assessing information related to the development of a wide range of security-related events, specific to the organization and its various operations that can adversely affect the security and safety of personnel and the profitability or reputation of the organization. Additionally, the CSO must logically determine the probability of these security-related incidents and develop appropriate preventive strategies consistent with sound business judgment and internal controls. The information to develop these assessments and preventive strategies may come from multiple sources, including organizational records, government and law enforcement agencies, news organizations, existing security bodies of knowledge, and elsewhere. The CSO should be capable of making the links between often disparate pieces of information from multiple sources to understand and assess their importance to the security of the enterprise. The CSO should understand and be familiar with the people skills and technological aids that will assist in this process, and possess both conceptual and critical thinking skills to prioritize risks and develop appropriate preventive strategies across the organization.

13.4 Organization Preparedness

The CSO is responsible and accountable for ensuring that the enterprise is prepared for the possibility of attack, catastrophic event, or related significant security incident (major fraud, product tampering, etc.). This will involve development and administration of training plans, programs, and exercises. A process of regular periodic review and evaluation of organizational readiness in the event of attack or event is a key responsibility of the CSO.

13.5 Incident Prevention

Another key responsibility of the CSO is analysis of information and the coordination of activities with persons inside and outside the organization to forestall and prevent attacks and catastrophic events. This implies the ability to successfully operate independently in fast-paced, matrix-management environments, requiring a high tolerance for ambiguity and positive political skills to drive programs and projects to completion.

The CSO also must identify and understand the nature of security risks in the business environment and the application of appropriate financial and managerial controls to mitigate those risks. This also will require the CSO to understand how and when to enlist the support of risk management, internal audit, controllers, outside resources, legal, human resources, and other staff functions also engaged in mitigating various risks to the business.

13.6 Securing People, Core Business, Information, and Reputation

Protection of the company's integrity, people, processes, and assets from harm and loss is also a key responsibility of the CSO. Though it is important to protect the financial and physical assets of the enterprise (cash, facilities, and equipment), the CSO also must be especially adept at countering the potential risks involved in the loss of intangibles (reputation), intellectual property, and trade secrets. People include management and directors, employees, customers, and others the organization has a duty to protect.

13.7 Incident Response, Management, and Recovery

In case of an incident of attack or catastrophe, the CSO will be responsible for coordinating efforts within the organization to restore critical systems and provide facilities needed by the organization to function.

The CSO will coordinate with internal and external resources to ensure adequate medical, financial, and emotional support assistance is provided to employees, customers, and others involved in a catastrophic event or an attack on the organization. The CSO will coordinate with local, state, federal, and international government agencies as required.

13.8 Investor Relations, Public Affairs, and Government Relations Coordination

The CSO must closely coordinate with those responsible for investor relations, public affairs, finance, human resources, operations, and government relations. Additionally, the CSO may be required to participate in the development of media interviews and testify before government regulatory agencies.

14.0 KEY COMPETENCIES

Generally, the CSO must be more strategic than tactical in orientation. Additionally, the CSO must have exceptionally strong business and interpersonal skills. The position requires a remarkably high degree of emotional maturity and the ability to calmly facilitate the appropriate resolution of difficult ethical and crisis situations. The ability to analyze, understand, and explain the value proposition of security initiatives to senior executives and Board members will be a key requirement of the position. It is likely the strategic, business, and interpersonal abilities of a CSO will be of greater importance than technical security skills, many of which are available through internal subject matter experts or external consultants.

The ability to communicate clearly, both orally and in writing, will be a very important competency. The interaction with senior executives and Board members means the person also must be comfortable in making presentations and fielding questions and challenges concerning the proposals and recommendations presented.

The CSO will need skills and competencies to accomplish the following:

- Relate to and communicate with senior executives, the Board of Directors, and its operating committees.
- Understand the strategic direction and goals of the business and how to intertwine security needs with the goals and objectives of the organization. This implies the ability to establish a vision for the global and individual business security programs and to build support for their implementation and ongoing development.

- Understand and assess the impact of changes in the areas of economics, geopolitics, organizational design and technology, and how they relate to potential threats and risks to the organization.
- Ensure security incidents and related ethical issues are investigated and resolved without further disrupting operations, and are conducted in a fair, objective manner in alignment with the organization's values and code of business conduct.
- Facilitate the use of traditional and advanced scenario planning techniques in assessing risks and threats to the organization.
- Understand how to successfully network and develop working relationships with key individuals in staff and line positions throughout the organization.
- Promote organizational learning and knowledge sharing through internal and external information resources in line with the culture of the organization.
- Be politically astute but not politically motivated.
- Be realistic and comprehend the need to assess the financial, employee, or customer implications of any plan or recommendation.
- Function as an integral part of the senior management team with regard to planning and capital expenditures.
- Develop organization-wide security awareness as appropriate for the business and the culture of the organization.

A description of the ideal CSO also should include the following personal characteristics:

- Strategic orientation with ability to act tactically, as required.
- Proven skills succeeding in a matrix-management environment.
- Global perspective, multi-cultural understanding and approach.
- Solid focus on detail, as required.
- Excellent conceptual and critical thinking skills.
- High integrity.
- Emotional maturity.
- Strong negotiator/facilitator and consensus builder.
- Sound understanding of process management principles.
- Ability to interact at all levels of the organization, and sensitivity to divisional/organizational management issues.
- Change agent.

15.0 EXPERIENCE

The contribution of prior experiences to the ability of the CSO to assess and determine success factors in the culture of the organization will be critical to the selection process. Demonstrated experience is key.

A broad and potentially diversified set of skills, education, and experience may be required depending upon the hiring organization's analysis of the position. The incumbent will be a change agent, able to be quickly recognized as a highly credible senior-level resource.

Depending upon the organization's market range and the job scope, some demonstration of international experience may be required, with added value being given to one or more language proficiencies.

The incumbent must have a range of experience that will permit the hiring organization to assess the challenges successfully addressed in prior experience to those likely to be confronted in the future. The desired candidate will be a seasoned manager with a collaborative outlook and a proven track record as a team player and business partner.

16.0 EDUCATION

This is a senior management position. As with its peers, there are significant expectations for education and experience. Advanced degrees are highly valued in all industries and represent the business connection that would likely enhance the CSO's credentials in many companies. Degrees in law, business administration, accounting and finance, security management, information systems management, or criminal justice also are well represented and should be considered, as should certifications in security and related disciplines.

The job-relatedness and benefit of education credentials must be balanced against the organization's culture. Typically, the quality, type of experience, and other directly related accomplishments will be a more compelling credential for the hiring organization.

17.0 COMPENSATION

The options for compensating this senior leadership position are far too wide, and the compensation practices of host businesses are too unique to be stated with confidence here. Recruiters with experience in this area, high-quality annual compensation analyses, and similar organizations who value a highly effective security program should be consulted for benchmarking.

(This page intentionally left blank.)

APPENDIX A

Model Position Description

Position Purpose

The incumbent serves as the executive responsible for the identification, development, implementation and management of the organization's [global][1] security strategies and programs.

Key Responsibilities

- In cooperation with the executive committee, directs the development of an effective strategy to mitigate risk, maintain continuity of operations, and safeguard the organization.

- Directs the domestic [and international] staff in identifying, developing, implementing, and maintaining security processes, practices, and policies throughout the organization to reduce risks, respond to incidents, and limit exposure and liability in all areas of information, financial, physical, personal, and reputational risk.

- Researches and deploys state-of-the-art technology solutions and innovative security management techniques to safeguard the organization's assets, including intellectual property. Establishes appropriate standards and associated risk controls.

- Develops relationships with high-level law enforcement [and international counterparts] to include in-country security [and international security agencies], intelligence, and private sector counterparts [worldwide].

- Through subordinate managers, coordinate and implement site security, operations, and activities to ensure protection of executives, managers, employees, customers, stakeholders, visitors, etc. and physical and information assets, while ensuring optimal use of personnel and equipment.

Key Skills and Competencies

- Leadership skills to provide direction to the management and professional staff within the organization.

- Ability to develop consensus within an organizational climate of diverse operational activities and often-conflicting regulations imposed by agencies with regulatory jurisdiction.

- Ability to effectively communicate within all levels of the organization, including briefing executive management and governance Board committees on status of security issues.

- Emotional maturity to understand impact and sensitivity of security issues.

Qualification Guidelines

- At least 3-5 years of direct experience in a significant leadership role. Demonstrated ability to develop and manage the functional capital and expense budget.
- Advanced degree or equivalent in an area of study relevant to this position and at least 10-15 years of experience in private sector corporate security or related public sector organization.
- [Demonstrated experience and exposure in the international security arena dealing with security-related issues.]

---

[1] Bracketed items are dictated by each organization's scope.